THE NUTTY JOKE BOOK

compiled
by
Charles Keller

illustrated
by
Jean-Claude Suarès

Prentice-Hall, Inc.
Englewood Cliffs, New Jersey

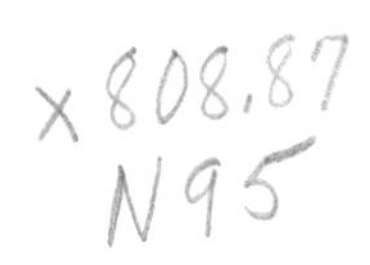

THE NUTTY JOKE BOOK

Printed in the United States of America

Prentice-Hall International, Inc., London
Prentice-Hall of Australia, Pty. Ltd., North Sydney
Prentice-Hall of Canada, Ltd., Toronto
Prentice-Hall of India Private Ltd., New Delhi
Prentice-Hall of Japan, Inc., Tokyo
Prentice-Hall of Southeast Asia Pte. Ltd., Singapore

Library of Congress Cataloging in Publication Data

Main entry under title:

The Nutty joke book.

SUMMARY: A collection of jokes about peanuts and nuts.

1. Wit and humor, Juvenile. [1. Peanuts—anecdotes, facetiae, satire, etc. 2. Nuts—anecdotes, facetiae, satire, and etc. 3. Joke book] I. Keller, Charles. 11. Suares, Jean-Claude.
PZ8.7.N87 817'.5'4080355 77-18897
ISBN 0-13-627737-3

To
Cheryl, Debbie,
Penny, Elsa and
Pat

What's the best way to catch a squirrel?
Climb a tree and act like a nut.

How can you tell if an elephant is in your refrigerator?
By the footprints in the peanut butter.

Why do elephants like peanuts?
Because marshmallows get stuck in their trunks.

How can you tell the difference between an elephant and a peanut?
Jump around on it for awhile. If you don't get any peanut butter it's an elephant.

When does a peanut
sound like a cashew?
When it sneezes.

What's yellow and lights up?
An electric peanut.

Why did Snoopy want to quit the comic strips?
He was tired of working for Peanuts.

What do you get when you cross an elephant and a jar of peanut butter?
Either an elephant that sticks to the roof of your mouth, or a jar of peanut butter with a long memory.

Did you hear about the baby who lived on peanuts for a week and gained ten pounds?
No. Whose baby was it?
An elephant's.

I once had to live on a jar of peanut butter for a week.
Weren't you afraid of falling off?

If you had ten bags of peanuts and you gave two to Jane and three to Mary, what would you have?
Two new girlfriends.

If I had ten bags of peanuts and gave you five, what would I have left?
I don't know; in school we learned with apples.

Do you know what a balanced diet is?
Sure—when you have a bag of peanuts in each hand.

If I had two wishes I'd turn everything into peanuts
and eat them all.
But you'd get sick.
Oh, I'd change myself into an elephant first.

If you had three bags of peanuts and I told you to
divide them with your brother, what would you
give him?
Do you mean my big brother or my little brother?

Fifty bags of peanuts, please.
All for you?
Don't be silly. I have two friends outside.

What's yellow and furry?
A rich peanut with a mink stole.

Why are peanuts yellow?
So you can tell them from watermelons.

What's small, yellow and has bucket seats?
A sports peanut.

What's yellow and has sixteen wheels?
A peanut on roller skates.

What's yellow and only comes out at night?
Vampeanut.

What's yellow, has a shell, and lays on its back?
A dead peanut.

How can you tell when there's an elephant in your peanut butter sandwich?
It's hard to lift.

What did the boy peanut say to the girl peanut?
Honey, I'm nuts about you.

What's red and white on the outside and yellow on the inside?
Campbell's cream of peanut soup.

How do you make a peanut squash?
Put it under a steamroller.

Why do elephants paint their toenails yellow?
I don't know. Why?
So they can hide in a bag of peanuts.
That's silly. I've never seen an elephant in a bag of peanuts.
See, it works.

What do you call a wild man who lives at the mouth of the Amazon River?
A Brazil nut.

What kind of medicine do you give a sick elephant?
Peanutcillin.

What do you do with a peanut that is one year old?
Wish it a happy birthday.

What do you get when you cross a peanut with a porcupine?
Splinters in your peanut butter.

What do you get when you cross a walnut and a banana?
I don't know, but it sure would be hard to peel.

What would you say if a man with a gun tried to rob your bag of peanuts?
Nuts to you.

What's yellow, has a shell, and lives on the bottom of the ocean?
A sunken peanut.

What would you do if you were attacked by an army of peanuts?

Shell them, of course.

What did one peanut say to the other peanut?
Nothing. Peanuts can't talk.

What does a peanut have in common with a bicycle?
Neither can play the piano.

What's the difference between a peanut and a mail box?
If you don't know, I wouldn't send you to mail a letter.

I call my boyfriend Candy Bar.
Because he's so sweet?
No, because he's half nuts.

I'm giving half my peanuts to the elephants.
That's nice.
Yes. I'm giving them the shells.

Doctor, I've had two peanuts in my ears since last year.
Why didn't you come to see me sooner?
I wasn't hungry until now.

Please give me some more peanuts.
If you eat any more peanuts, you'll burst.
Pass the peanuts and get out of the way.

What's worse than a worm in a peanut?
Half a worm.

Is it okay to eat peanuts with your fingers?
No, peanuts should be eaten separately.

What's the best thing to put into a peanut butter sandwich?
Your teeth.

How can you divide ten peanuts between nine people?
Make peanut butter.

What looks like half a peanut?
The other half.

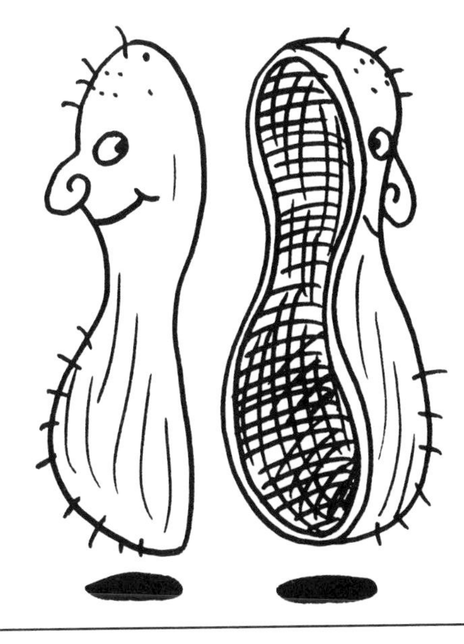

What's red and eats peanuts?
An embarrassed elephant.

Why do elephants eat peanuts?
They're saltier than prunes.

Why do monkeys prefer peanuts to caviar?
They're easier to get at the ball park.

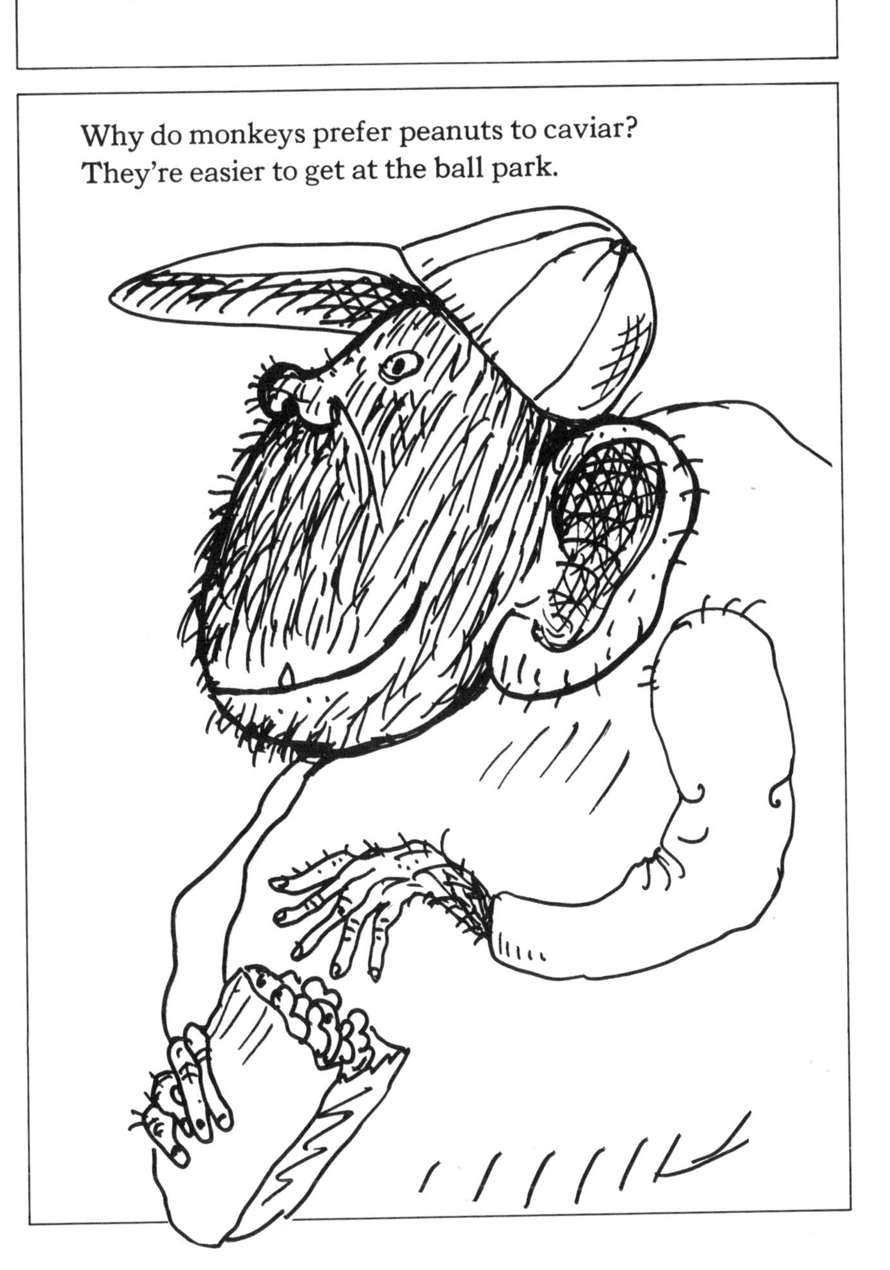

Why do elephants jump out of trees?
To squash peanuts.

What do elephants eat beside peanuts?
Canned elephant food.

How can you tell if an elephant is in the back seat of your car?
You can smell the peanuts on his breath.

What's tall and yellow?
The Empire State Peanut.

Why is there usually more than one peanut in a shell?
To keep the other one company.

How can you tell a peanut from spaghetti?
A peanut doesn't slip off the end of your fork.

What's the difference between a peanut and an elephant?
An elephant is grey.

When can you see an elephant in a bag of peanuts?
Never. They come in boxes of Cracker Jacks.

Why did the farmer go over his field with a steamroller?
He wanted to raise peanut butter.

What is the best way to keep peanuts?
Don't give them away.

If you raise peanuts in dry weather, what do you raise in wet weather?
An umbrella.

What's yellow and goes beep-beep?
A peanut in a traffic jam.

I love you, I love you,
I love you so well.
If I had a peanut,
I'd give you the shell.

Now I lay me down to sleep,
A bag of peanuts at my feet.
If I should die before I wake,
You'll know I died of a bellyache.

A peanut sat on a railroad track,
His heart was all a-flutter.
Around the bend came a railroad train;
Toot, toot, peanut butter!

Where were the first peanuts grown?
In the ground.

Did you eat all the pistachio nuts?
I didn't touch one.
But there's only one left.
That's the one I didn't touch.

I et seven peanuts.
You mean ate, don't you?
Maybe it was eight I et.

Did you hear the joke about the guy who had peanuts in his ears?
No.
Neither did he.

How much are these peanuts?
75¢ a pound.
Did you raise them yourself?
Yes. Yesterday they were 60¢ a pound.

Why did the peanut cross the road?
To get to the shell station.

How can you tell which peanuts are left-handed?
Place a large bag of peanuts in front of someone and ask them to eat as many as they can. The remaining ones are left.

Why don't peanuts laugh?
With all these terrible peanut jokes going around, how can they?

I sent my son for five pounds of peanuts and you only gave him three pounds. Are you sure your scale is correct?

My scale is accurate; have you weighed your son?

Guess how many peanuts I have in my hand.

I don't want to guess.

Go ahead and guess.

I don't want to guess.

If you guess you can have them both.

Want a peanut?
No, they make you fat.
What makes you say that?
Have you ever seen a skinny elephant?

What's yellow and points north?
A magnetic peanut.

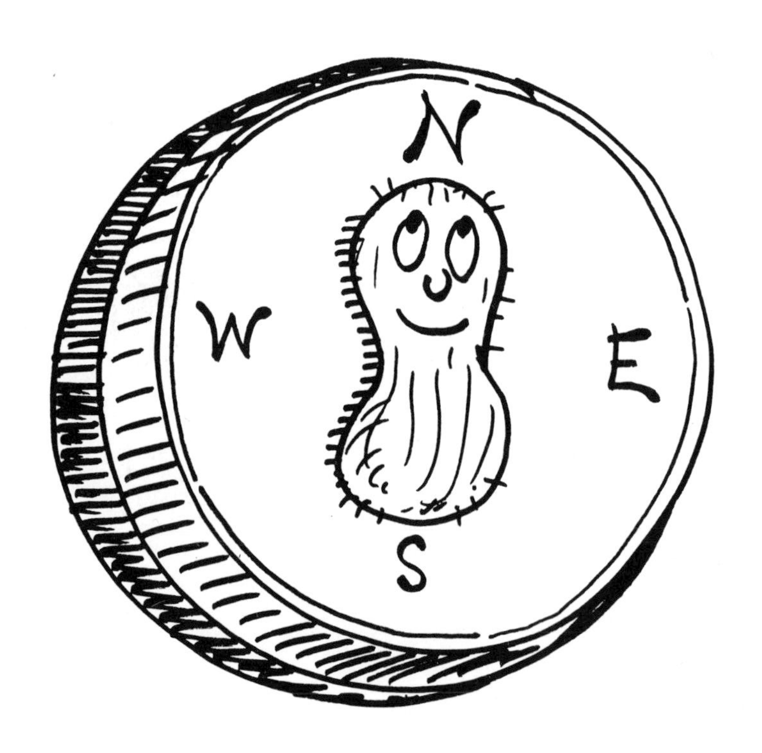

What's yellow, wears a mask and rides a horse?
The Lone Peanut.

What's yellow and black and blue?
A bruised peanut.

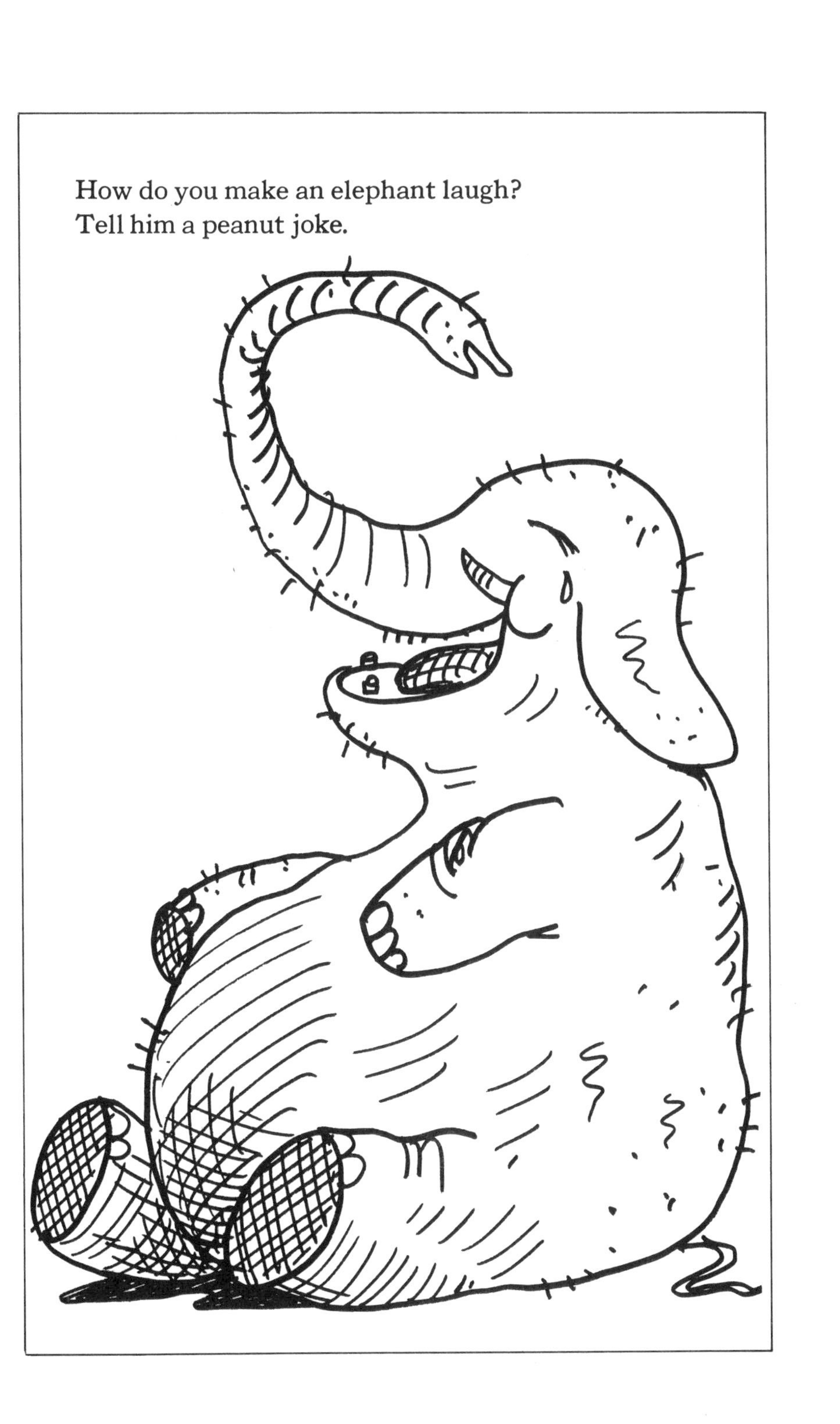
How do you make an elephant laugh?
Tell him a peanut joke.

What's yellow and seldom rings?
An unlisted peanut.

Why don't peanuts like to lend money?
They're tired of people putting the bite on them.

How do you say peanut in Spanish?
Peanut in Spanish.

If a peanut worker was six feet tall and wore a size nine shoe, what would he weigh?
Peanuts, of course.

What's yellow and hides in caves?
A peanut that owes money.

What's yellow and goes, "snap, crackle, pop?"
A peanut with a short circuit.

If corn has ears and potatoes have eyes, what do peanuts have?

Each other.

Why do elephants eat so many peanuts?
So they can send in the wrappers for prizes.

Did you hear the joke about the old bag of peanuts?
No.
Never mind, it's stale.

Will you join me in a bag of peanuts?
No, I don't think we'd both fit.

How do you know when you've eaten too many peanuts?
I don't know.
When the person sitting next to you is you.

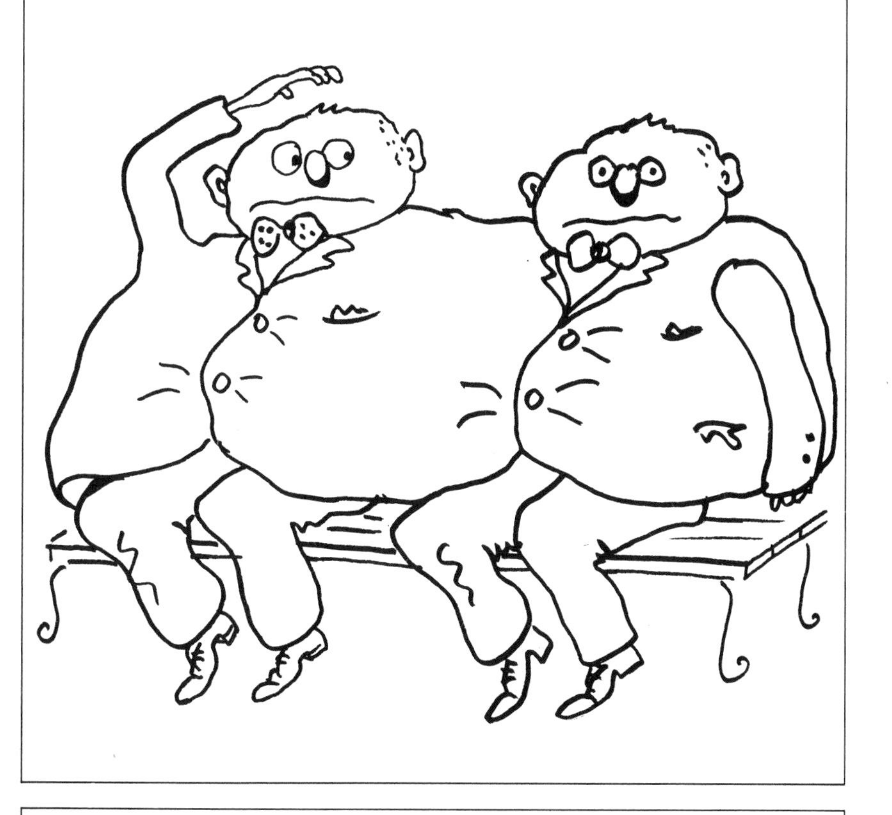

Do you like peanuts?
No.
Good. Will you hold this bag while I get a drink?

How come a nut like you was missed in the peanut harvest season?

The same way they missed you on Thanksgiving.

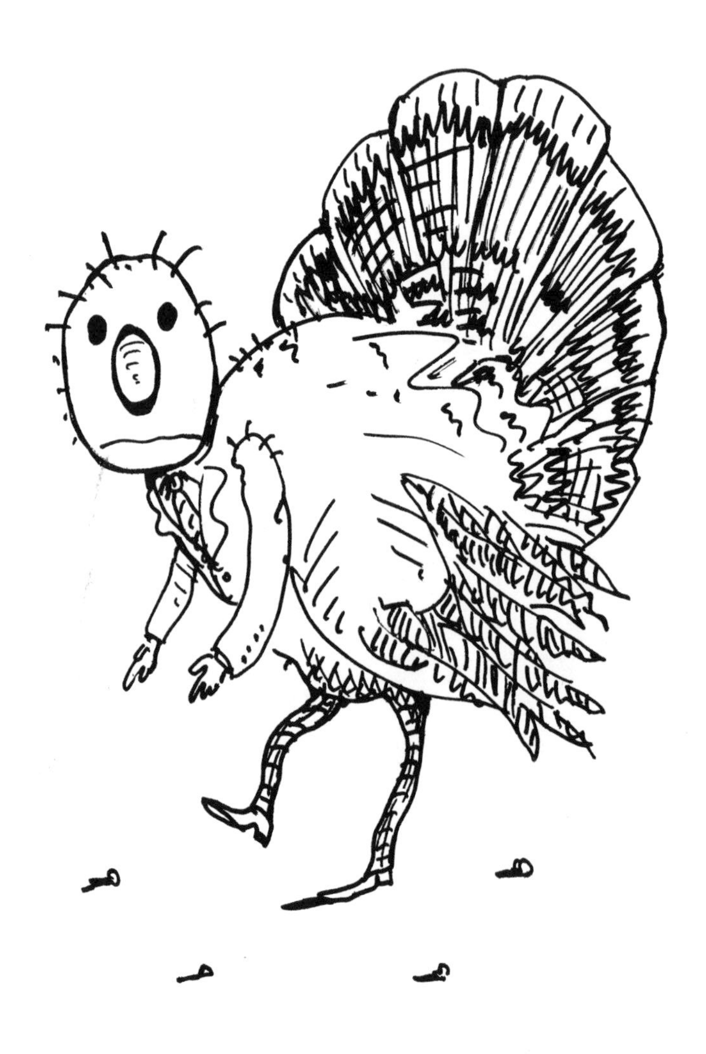

What's yellow and fights crime?
The Bionic Peanut.

Where do you find peanuts?
It depends on where you lost them.

What do you get when you cross a peanut with an onion?
A peanut with watery eyes.

What's grey, has a trunk and weighs ten pounds?
An elephant that's allergic to peanuts.

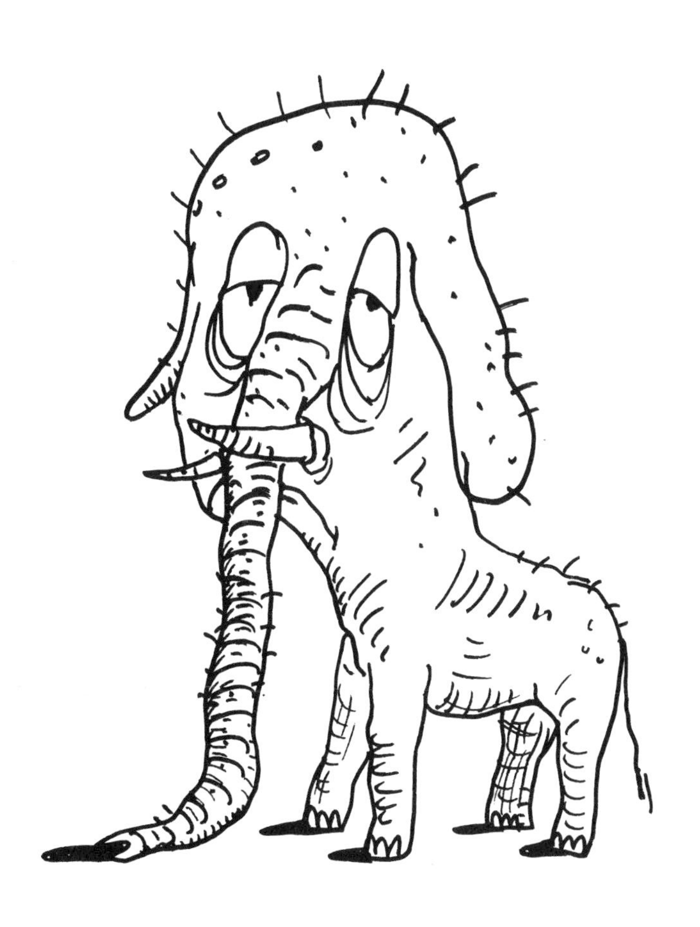

What nut would you hang a picture on?
A walnut.

What nut is found by the sea?
A beechnut.

What's grey and has brown feet?
An elephant that makes his own peanut butter.

How do you make a peanut sundae?
You start by getting it ready Fridae and Saturdae.

How many sides does a peanut have?
Two, an outside and an inside.

What's yellow, then purple, then yellow, then purple?

A peanut that moonlights as a grape.

That's funny. You have peanuts growing out of your head.
I know. I planted radishes.

Can you eat peanuts with fingers.
No, peanuts don't have fingers.

Mom, what are we having for lunch?
Oh, hundreds of things.
Good. What are they?
Peanuts.

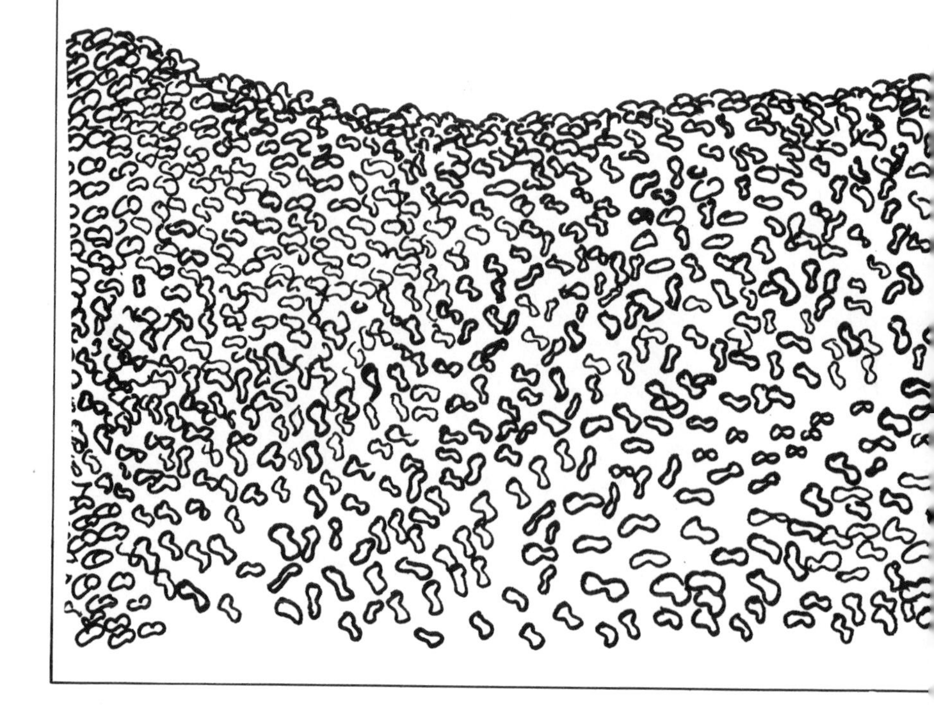

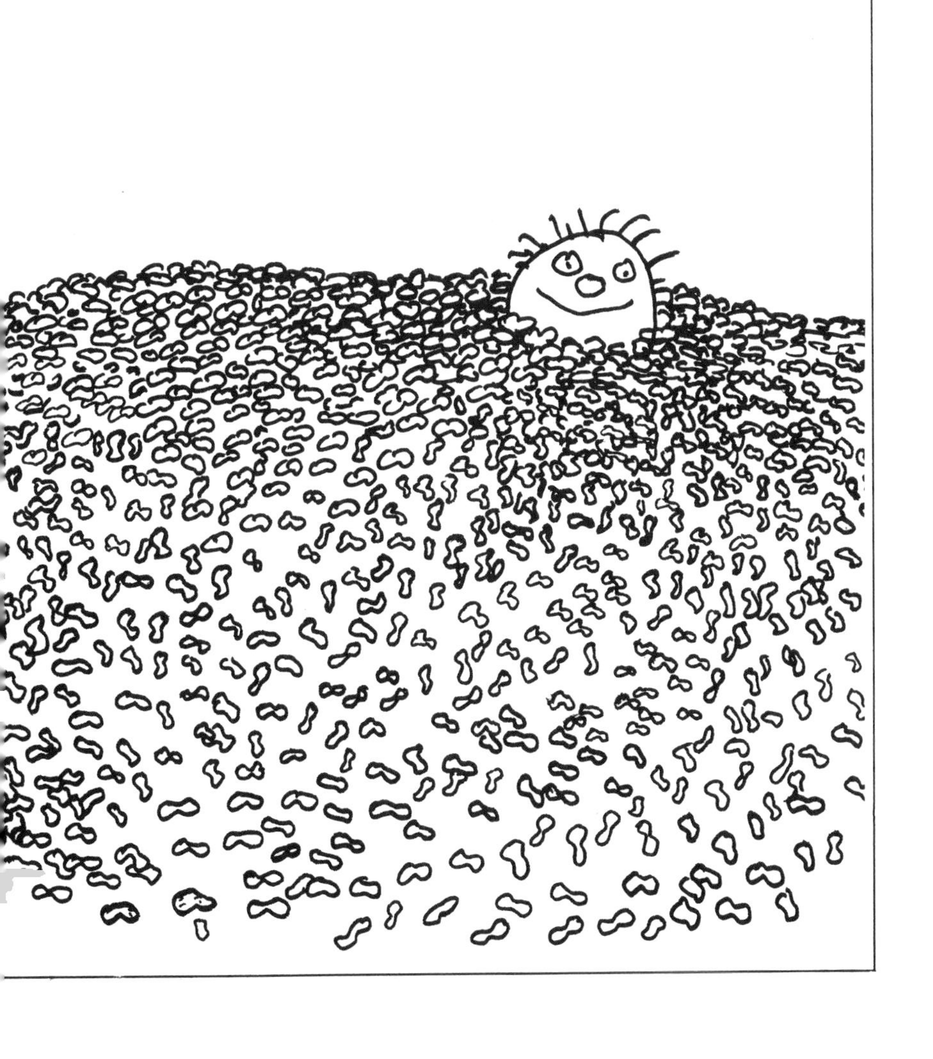

What is the elephants' best seller?
Peanuts.

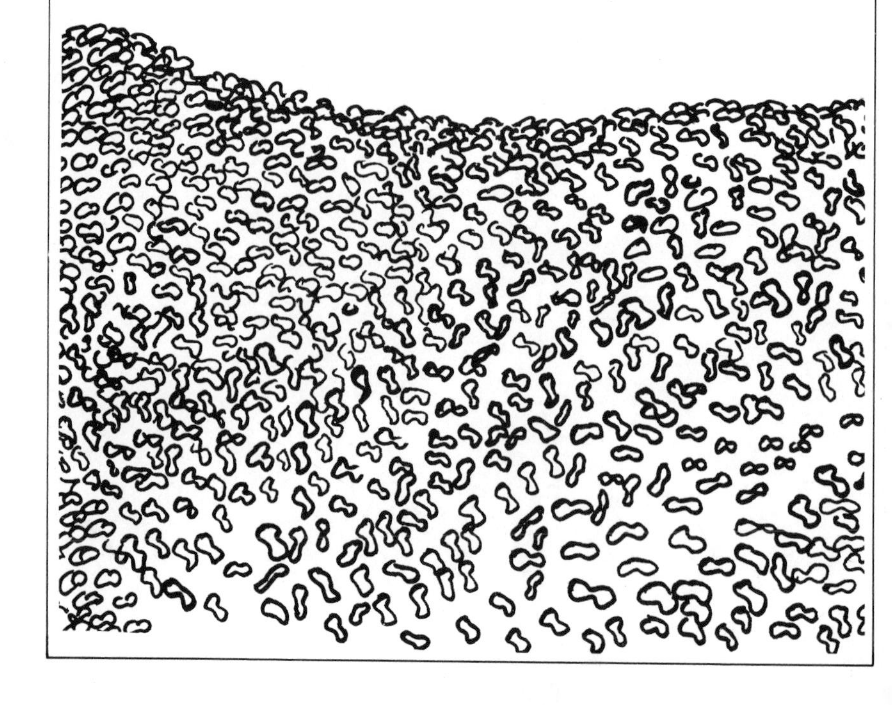